Dodging here and dodging there,
they have to miss shapes everywhere!

So off they go — their engines burn —
passing spheres that spin and turn.

But look at what lies straight ahead —
the zone of cones all spaceships dread.

They're safe — but on the other side
they reach a place where monsters hide!

And still to come are cubes galore,
and other box shapes — more and more!

They slide along a great flat face,
so glad they're nearly back to base.

After such a tiring test,
the spaceships both deserve a rest.